ELEVATE YOUR STANDARDS

Building an Operations Manual for your Uncrewed Aircraft Systems Company

I dedicate this book to my cherished family
and in memory of my Papa.
"Going downtown" always meant you were going flying.

ELEVATE YOUR STANDARDS

Building an Operations Manual for Your Uncrewed Aircraft Systems Company

Desiree Ekstein

ELEVATE YOUR STANDARDS

Building an Operations Manual for your Uncrewed Aircraft Systems Company

Desiree Ekstein
Building an Operations Manual for your Uncrewed Aircraft Systems Company (Rev. ed.)

First published: December 2021
Revised Edition: October 2024

ISBN: 978-0-578-32622-1(paperback)
ISBN: 978-0-578-32621-4 (Kindle)

Cover design: Gloria Lopez

Contents

This page intentionally left blank.

Acknowledgments

"With God all things are possible" - Mathew 19:26

I thank the people who have helped me with my life goals and writing this book with tremendous gratitude. My friends and family were there along to way to "elevate" and encourage me to share my passion.

Thank you to my family for all your love and support for allowing me to follow my dreams. You have helped me get where I want to be and often went with me on the road trips to get there.

Carolyn White, for all your time and knowledge you have spent collaborating with me to bring this manuscript together. Your guidance was invaluable to me as I came down to the final touches.

Eric Hanscom, thank you for always answering my emails and providing a direction to search for my legal questions and information.

Louise Jupp, I am deeply indebted to you for being a shining example, providing your input, knowledge, and expertise in the field of education, editing, and writing. I truly appreciate the hours of consultation and your guidance through the intricacies of the manuscript and publication.

Pharns Genece, I cannot thank you enough for always finding time to help me and giving me advice from a higher perspective. Through your training and military service, you set an example of professionalism and know the importance of standards and protocols within a company.

Sharon Rossmark, I am incredibly grateful for opening all the doors and encouraging me to take the first step. For guiding me, correcting me, and showing support. You have allowed me to bounce all my ideas by you, and you have provided opportunities for me to share my passion and training. This book would not have been possible without your support and guidance.

Timm Keeley for being an inspiration and "teaching the teacher." Rest in peace. I wish you were here to see the final product.

This page intentionally left blank.

Introduction

Elevate your Standards. Building an operations manual for your uncrewed aircraft systems company.

Hello, I'm Desiree Ekstein, also known as Drone Diva® Desi. Thank you for choosing "Elevate Your Standards" for your uncrewed aircraft systems safety guidance. You've made an important choice in getting started with elevating your company safety standards. If you have an uncrewed aircraft operation, you know the importance of safety and want to elevate your standards of professionalism, this book is for you.

I love when I talk with aspiring operators that are just starting their new journey with the uncrewed aircraft, they've passed the exam and ask, "What's next?"

I want you to be a success. I will share some of my experiences and the safety standards I have learned through years of flying. This information is based on the Federal Aviation Administration (FAA) regulations at the time of this writing, December 2021[1] and updated in October 2024.

Ok, so by now, you are wondering why "uncrewed aircraft"? While working with professionals in the industry and the general public, I have found the use of the term UAV (Unmanned Aerial Vehicle), UAS (Unmanned Aircraft System), sUAS (small Unmanned Aircraft System), and Drone (aircraft without a human pilot, crew, or passengers on board) all frequently used. The term "drone" has become commonly used but not always considered professional. Generally, when working with a client or speaking with the public, "drone" is identifiable. Initially, I planned to use "drone" in this book. As the industry evolves, new terminology is also evolving, and "uncrewed" is becoming more widely used. Throughout this book, I will use aircraft or uncrewed aircraft. If we are going to be discussing elevating standards and professionalism, it is best to use a term that will elevate and distinguish you as a professional.

Safety Management System and Standard Operating Procedures are an integral part of formalizing operations within a company. Every uncrewed aircraft service provider has unique operating circumstances. Safety Management Systems encompasses safety within all aspects of your company from the top down. Standard Operating Procedures (SOPs) are a part of the Safety Management System (SMS). While SOPs are an ongoing process within a company, getting started is the hardest part. SOPs define company policies, training,

---

[1] https://www.ecfr.gov/current/title-14/chapter-I/subchapter-F/part-107#part-107

emergency procedures, asset maintenance, mission directives, and team member responsibilities. Set standards will assure compliance with industry codes, data transfer, and FAA regulations through documentation. They provide effective communication between team members and eliminate confusion. A company will operate more smoothly and efficiently when they have a set of standards in place.

No two companies will operate the same, but a base for safety should be implemented. Further attention should be given to the specifics of the individual priorities within a company and their needs. This document is in no way complete and should not be used as the only resource for safety and risk mitigation. Additional sources for safety and training are ANSI[2], AUVSI TOP[3], JARUS SORA[4], and NIST[5]. SMS and SOPs are an ongoing process in your workflow. As positive advancements and experiences in your operations occur, your SOPs should be developed and used as a training tool.

This book is subdivided into three sections, company policy, standard procedures, and emergency procedures. There is also an "additional information" area for you to add any details specific to your operation. It is recommended that the suggested checklists be supplied to all team members to follow.

1= Company policy
2= Standard procedures
3= Emergency procedures

An excellent place to start building company standards is by defining you and your company. What are your goals, and what is your purpose for the company? What are the objectives you are going to satisfy?

I've identified ideas of what to include and have provided a starting point. There is a section for you to add your business objectives and on-site procedures. I recommend making notes on each job for the specifics you want to include.

Let's dive right in.

---

[2] https://www.ansi.org/
[3] https://www.auvsi.org/topoperator
[4] http://jarus-rpas.org/sites/jarus-rpas.org/files/jar_doc_06_jarus_sora_v2.0.pdf
[5] https://www.nist.gov/system/files/documents/2019/08/21/nist-astm-nfpa_standard_test_methods_for_suas_-_maneuvering_and_payload_functionality_overiew_v2019-08-20v2.pdf

Company policy

Policy

Objectives

As an uncrewed aircraft company, SOPs are developed to provide instructions and guidance to a team for a specific mission or operation. These training materials are expected to be performed as needed per job and testing. The purpose of an SOP is to provide detailed written instructions to achieve uniformity of the performance of a specific function. The SOP will document the necessary preparation and activities in the pre-flight stage, during operational flight, post-flight activities, safety procedures, and documentation requirements. This will include all safety checks, pre-flight equipment preparation, detailed description of location and flight exercises, roles and responsibilities of participants, regulatory and environmental documentation and reporting, and post-flight activities and documentation. Test and study materials are provided by the FAA, industry books, internet, hands-on, and AUVSI TOP, ANSI, NIST, among others.

Your company focus is:

____

____

____

____

Mission Statement

Provide a mission statement for your company. A mission statement describes what our company objectives are, your purpose, and who is your expected clientele. What are your overall goals for the company? What is the focus of your company, your clientele, and your team members?

> Example: *On The Go Video.biz is focused on UAS safety and education as well as aerial imagery services. We provide data from marketing to 3D images. On The Go Video.biz is committed to promoting the aerial industry as a whole through training, education, and providing valuable data, emphasizing the activities required for safe operations.*

Your company mission statement:

____

____

____

____

Code Of Conduct

You and your team members are always a representative of the company. From the time you first talk with the client until the data is delivered. As company representatives, all activities are a reflection of your company and the team member. This includes appearance, mannerisms, and speech. Therefore, any violations of the code of conduct will reflect poorly on the individual, the team, the UAS industry, and your company.

Safety of any mission is the primary goal, and the Code of Conduct will reflect this focus. The importance of safety cannot be underestimated because any violations of the code of conduct will reflect poorly on the work ethics of your company and potentially jeopardize future missions.

The CODE of CONDUCT of the SOP includes (but is not limited to):

- All activities and decisions should be made with SAFETY in mind and with respect to the company's "chain of command."
- All flights are subject to proper pre-flight checks and hazard/weather/advisory checks.
- Uncrewed aircraft flights should be within defined boundaries and away from any people, neighbors, and private areas.
- Team member behavior should be conducted with courtesy and respect.
- Flights should focus on specific tasks as defined by the team lead.
- Each team member should take part in their defined task.
- Team members will dress according to company standards. This will include proper shoes, pants, and a shirt. Steel-toed boots and a hard hat will be worn when required.

Additional notes and information:

Company Insurance

When I first started flying uncrewed aircraft, insurance was a tricky one. I had to go through an aviation company that would also provide insurance for uncrewed aircraft. Lucky for us, several insurance companies are geared towards this aircraft now, and we have more options available. Insurance will elevate your professionalism, protect you and your client. Identify your insurance requirements. When choosing an insurance company, consider all the options and what will fit your needs. "On-demand" insurance is available and may be an option for you. Take into consideration general liability for the company as a whole. Will the carrier provide an "Additional Insured Rider or Endorsement? Are your aircraft and ground station equipment covered? These are things you should talk over with a reputable insurance agent/broker.

Insurance provider information. Insurance options to consider including are general liability, professional liability, and aircraft coverage. Always maintain a copy of your insurance on file with your company records. A digital and paper copy should be with you and available on the job site should you be asked for documentation.

Additional notes and information:

Human Resource Department (HR)

HR is outside the scope of this document, and legal guidance should be obtained.

If you are just getting started, you may not have a crew or staff, but it may be your goal to have a large operation with a full-scale team. Alternatively, you may have a team of independent contractors. If so, you will want to keep documentation regarding the team.

Additional notes and information:

HR Department

HR Department							
Date Hire	First Name	Last Name	DL #	RPIC #	SS #	I9	W2

Figure 1 - HR department.

Pilot Training and Information

Keeping a log of pilot training and current information is critical to elevating your standards.

Pilot training must be documented and logged. Additionally, newly acquired skills should be annotated for each team member as needed. Added certifications would include but are not limited to Cardiopulmonary resuscitation (CPR), Thermography, Occupational Safety and Health Administration (OSHA).[6]

A checklist specific to the pilot training evaluation and flight maneuvers is provided in the Pilot Evaluation Training Checklist section.

Additional notes and information:

Pilot Training Log

Pilot Training Log							
Pilot	Date Hire	Cert Renew Date	Cert Number	Cert Date	Training Date	Training Level	Additional Cert's

Figure 2 - Pilot training log.

[6] https://www.osha.gov/

Pilot Flight Record

Keeping a flight record from the start is a lesson I learned the hard way. I flew for many years before I started tracking my flight hours. I suggest you start right from the beginning. You may find companies will only look at the hours after you obtain your Remote Pilot Certificate. Your insurance company may look at your total flight hours, and that could save you money on premiums.

A pilot's log has many benefits and must be available for the FAA inspection upon request. I recommend keeping a paper and electronic copy. A log can be maintained via an online service, by hand, or by creating a personal document or spreadsheet. The log will document all pilot information as written as proof of a pilot's flight time and proficiency. This document can also be of value for insurance purposes, verification for employment, and assists with FAA waiver approvals. This can be done through an online service such as AirData – Flight Data Analysis[7]. There are also many books and online resources.

Pilot Flight Log

Pilot Flight Log						
Date	Aircraft Make/Model	Flight Location	Flight Purpose	Qty Batteries Used	Flight Duration	Notes

Figure 3 - Pilot flight log.

[7] https://airdata.com/

Fleet Management

Documenting the records of your fleet will assist you with your business organization. As your company grows, having a log of your inventory will assist with organization. Log of your fleet, the date of purchase, make/model, and where the asset was purchased, and the pertinent FAA information must be recorded. A documented log of the company aircraft fleet will also include documentation of the maintenance, firmware updates, and any necessary repairs performed.

A spreadsheet for individual aircraft maintenance has been provided in the section Aircraft Maintenance Log.

Documentation and registration numbers must be kept with the aircraft.

Each commercially used aircraft must be registered with its own unique FAA number. This is done through the FAA DroneZone[8] website. The FAA registration number must be visible on the outside of the aircraft. The marking can be done by engraving, permanent label, or permanent marker. Registration is $5.00 and valid for three (3) years.

The FAA requires an aircraft to broadcast a Remote Identification number. This number is required during registration for commercially flown aircraft. More information about Remote Identification can be found in the following sections.

See links and resources page for links to the FAA website for registration.
https://faadronezone-access.faa.gov/#/

Additional notes and information:

[8]https://faadronezone-access.faa.gov/#/

Remote Identification (Remote ID)

One of the most significant advancements in drone regulation is the implementation of Remote Identification (Remote ID, or RID).

As of September 16, 2023, all Part 107 pilots, as well as recreational drone pilots, must comply with Remote ID regulations. There are a few exceptions, such as flying in an FAA Recognized Identification Area (FRIA) zones or operating drones under 250 grams that do not require FAA registration. **If your drone is used for commercial purposes must be registered and Remote ID is required.**

Remote Identification is often referred to as a "digital license plate" for drones. It is a system that allows the identification of a drone in real-time, providing information such as the drone's location, the pilot's location, and other key operational details. Drones must transmit their identity and location data during flight.

Types of Remote ID Compliance

There are three ways drone pilots can comply with the Remote ID requirements.

1. **Standard Remote ID**: Drones are equipped by the manufacturer's built-in capabilities to broadcast Remote ID information. This will broadcast all the pertinent information including the drone's serial number, location, altitude, velocity, and the control station's location.

2. **Broadcast Module**: For drones that do not have Remote ID built-in, pilots can attach a separate broadcast module. This module transmits the drone's take-off position and necessary Remote ID information during the flight.

3. **FAA-Recognized Identification Area (FRIA)**: Pilots flying within a FRIA do not need to broadcast Remote ID information. These areas are typically associated with community-based organizations or educational institutions where drones can be flown without Remote ID.

The Remote ID number can be found in the drone software, often under the "About" tab. If you are unsure, check the UAS Declaration of Compliance https://uasdoc.faa.gov/listDocs to determine if your drone or broadcast module is on an FAA-accepted Remote ID Declaration of Compliance (DOC) list. When utilizing the website for Remote ID, ensure you are searching with the proper RID filter.

Category Drones

Operations Over People (OOP) and Operations Over Moving Vehicles (OMV) Drone Categories

Remote ID opens doors to the possibility of OOP and OMV. Your drone must be Remote ID compliant to operate over human beings or over moving vehicles. This doesn't mean if you have Remote ID, you are good to go fly over human beings. There are a few more details to address.

You must identify what category your drone is and if the drone is compliant for OOP with the FAA.

The FAA classifies drones into four categories based on their operation and potential risk to people and property. Category 1, Category 2, Category 3, and Category 4. Each category has specific details about size, payload, risk and use case.

Full details on OOP or OMV can be found in the Code of Federal Regulations Part 107 subpart D. This section also defines category specifics of the drones.
https://www.ecfr.gov/current/title-14/chapter-I/subchapter-F/part-107/subpart-D

But wait… there's more.
Once you have determined what category your drone is in, your next step is to confirm the drone has been approved by the FAA. Utilize the same FAA DOC website and search with the OOP filter. Not all drones are compliant.
https://www.faa.gov/uas/commercial_operators/operations_over_people

Additional notes and information:

Fleet Registration and Purchase

Fleet Reg. and Purch.							
Date Purchased	Aircraft Make	Aircraft Model	Purchased From	SN # & RID #	FAA #	Date FAA #	Date Renew FAA #

Figure 4 - Fleet registration and purchase.

Fleet Maintenance

Fleet Maintenance							
Date	Aircraft ID	Aircraft Number	Personnel	Maintenance Performed	Firmware #	Notes	Result

Figure 5 - Fleet maintenance record.

Certificates of Authorizations or Waivers

The FAA must authorize approval to operate in controlled airspace.
A certificate of authorization (COA) or a certificate of waiver (COW) is a certificate granted by the FAA for special use. Generally, a COA or a COW is used for a specific purpose or long-term use.

Other airspace authorizations are obtained through the FAA DroneZone or the Low Altitude Authorization and Notification Capability (LAANC)[9] applications. There are various providers for the LAANC application. LAANC approvals can be instantaneous. However, you should not rely on a quick response. Approval could take longer based on the local Air Traffic Control (ATC). You would not want to plan for a job and find the approval has taken longer and delayed your mission. If the LAANC application has a software glitch and becomes inoperable, your job could be delayed or canceled. Some applications require 48-hour notice. If airspace authorization is not available through a LAANC application, approval must be obtained through the DroneZone portal. The DroneZone portal could take 90 days.

DroneZone portal:
https://faadronezone.faa.gov/

Information regarding COA/COW
https://www.faa.gov/about/office_org/headquarters_offices/ato/service_units/systemops/aaim/organizations/uas/coa/

Additional notes and information:

[9] https://www.faa.gov/uas/programs_partnerships/data_exchange/

COA / COW

COA / COW						
Date Obtained	Date Expires	COW/COA Name / Area	COW/COA #	Filed By	FAA Contact	Notes

Figure 6 - Certificate of authorizations (COA/COW).

Training

Training, practice, and knowledge of the current regulation are a continuous process. You know the saying "use it or lose it"? It is so important to keep up with your flight skills and the requirements of the FAA. Dedicate specific flight times to ensure you feel comfortable on the sticks.

All pilots are required to be current with the FAA Remote Pilot Certificate to operate for commercial purposes.

As standards, pilots and visual observers are to be knowledgeable of FAA regulations. A practical flight assessment (PFA) will be utilized to demonstrate flight skills.

Pilots are trained to inspect the aircraft before and after each mission.

Your company may also require pilots, visual observers, and sensor operators to provide pilot flight logs and/or background information regarding previous training.

When necessary, re-current testing, PFA, and additional training will be required to be mission-specific compliant.

Additional notes and information:

Remote Pilot Certificate

FAA Part 107 Remote Pilot Certificate training is required for uncrewed aircraft commercial operations in the United States. As part of standard SOPs, knowledge of the current FAA regulations is required. Obtaining the Remote Pilot Certificate training is beyond the scope of this book.

FAA Part 107 Remote Pilot Certificate study materials will cover FAA Regulations, National Air System (NAS), Weather, Load Factors, and Human Factors.

Additional notes and information:

Aircraft Inspection

Elevating your standards includes doing a thorough examination of the aircraft and equipment before and after each mission. The aircraft's airworthiness is essential to safe flights. When there are noticeable damages or repairs required, they will be noted on the aircraft maintenance log. Discontinue flight missions until the aircraft is in proper working condition.

Inspect the aircraft as pre-flight preparation

- All sides of the aircraft.
- Battery prongs on the aircraft and batteries.
- Gimbals.
- Motors.
- Propellers.
- Screws and attachments.
- Wiring.

Remote control as pre-flight preparation

- Antenna.
- Buttons.
- Controls.
- Switches.
- Thumbsticks.

Additional notes and information:

Flight Training

When first starting your journey with your aircraft, you may not know where to start developing the skills that build confidence. Start with the Level 1 basics. By mastering basic maneuvers, such as a box, you will develop the muscle memory that is needed to have an automated response to the aircraft's input. Once you feel comfortable with the basic movements, then push yourself, continue the maneuvers with the aircraft nose facing various directions.

Level 2 adds new maneuvers to enhance precision and multiple control functions. Level 2 also includes maneuvers at various elevations and distances. Practice these same maneuvers in “sport” or Attitude (“Atti”) mode. Using these modes will remove the Global Positioning System (GPS) function on your aircraft, so fly with caution.

Level 3 training is industry and software-specific and is outside the scope of this book. It is a deeper dive into training requirements for specific industry missions, such as construction, inspections for infrastructure, and Search and Rescue (SAR).

Flight training is necessary to ensure a pilot has been adequately trained for the performance and skill required for the operation. A pilot must maintain precision flight with controlled stick movements. Basic and advanced flight training maneuvers will be practiced and implemented. AUVSI and NIST flight and testing procedures are also used as a guide to test skill levels. A Practical Flight Assessment (PFA) is given to assess pilot skills and to track improvements.
ANSI,[10] ASTM,[11] AUVSI TOP,[12] and NIST[13] have resources available for additional flight training information.

[10] https://www.ansi.org/
[11] https://www.astm.org/
[12] https://www.auvsi.org/topoperator
[13] https://www.nist.gov/ctl/pscr/research-portfolios/uncrewed-aircraft-systems

Pilot Evaluation Training Checklist

Level 1

Basic flight maneuvers and stick skill – Surface (SFC) to 200' AGL. Demonstrate proficient flight skills with launch/land, ascend (climb), descend, pitch, yaw, roll, and aircraft nose facing away, leading direction, and facing the pilot.

Level 1					
Pilot	Flight Maneuver	Complete	Evaluation	Date	Notes
	Launch / Land				
	Hover over cone				
	"T"				
	Box (square)				
	Slalom cones (weaving)				
	Figure 8				
	Glideslope				
	POI / Pan 360° rotation				

Level 2

Controlled exercises – SFC – 400’ AGL. Additional maneuvers.

Level 2					
Pilot	Flight Maneuver	Complete	Evaluation	Date	Notes
	Level 1 at higher AGL				
	Level 1 "Atti" mode				
	POI / Pan 360° rotation				
	Flight near objects				
	Autonomous flight				
	Camera skills				
	Hover, tilt, rotate				
	Glideslope w/ Camera skill				
	Distant launch/land				

Level 3

Industry and Software Specific

Level 3					
Pilot	Flight Maneuver	Complete	Evaluation	Date	Notes
	Agriculture				
	Construction				
	Renewable energy				
	SAR				
	Analytics				
	Mapping				
	Modeling				
	Photography				

Figure 7 - Pilot evaluation training checklist.

Flight Levels Checklist

Level 1

Basic flight maneuvers and stick skill – SFC to 200’ AGL.
Demonstrate proficient flight skills with launch/land, ascend (climb), descend, pitch, yaw, roll, and aircraft nose away, leading direction, and facing pilot.

- ☐ Launch / land
- ☐ Hover over cone
- ☐ “T”
- ☐ Box (square)
- ☐ Slalom cones (weaving)
- ☐ Figure 8
- ☐ Glideslope to various elevations (bases on flight location)
- ☐ POI / Pan 360° rotation

Level 2

Controlled exercises – SFC – 400’ AGL. Additional Maneuvers.

- ☐ Level 1 at higher AGL
- ☐ Level 1 in “Atti” mode
- ☐ Camera skills:
    - Glideslope camera tilt
    - Hover, tilt, rotate exercise with fixed points
    - POI with camera focus on point
- ☐ POI / Pan 360° rotation
- ☐ Autonomous software flights, i.e., POI, Follow me, Tripod mode
- ☐ Flight skill around, near, and through objects

Level 3

Industry-and Software-specific.

- ☐ Industry-specific (one or more)
    - Agriculture
    - Construction
    - Renewable energy
    - SAR
- ☐ Software
    - Analytics
    - Mapping
    - Modeling
    - Photography

Figure 8 - Flight levels checklist.

Night Operations Safety

Night operations are included with the FAA Part 107 Remote Pilot Certificate. When flying at night, new hazards and concerns become apparent. It is one thing to pass the FAA exam, and it is another to fly a mission at night. For night operations, there are regulatory requirements and safety concerns that apply to flying at night. Know the anatomy of the eye and how it will affect your night vision. Practice, practice, practice until you feel comfortable with the night flight. The first time you experience flying at night can be exhilarating, but a night flight will cause some added tension. Elevate your standards when flying at night by taking extra precautions. Do pre-flight inspections. Inspect the site during daylight hours. Utilize shorter flight times and proper visual scanning techniques.

Night flight considerations:

- Forms of communication.
- Increase your visual observers.
- Lighting equipment for ground station and aircraft.
- Navigational aids.
- Pre-flight planning and on-site setup.
- Understand and know the effects of visual illusions.

I highly recommend taking a night operations course. There is so much knowledge to gain by learning from an instructor skilled with night operations.

Additional notes and information:

____

____

____

____

The FAA has an entire chapter in the Airplane Flying Handbook[14] dedicated to night operations. Build your SOPs upon mitigating risk and addressing the performance-based standards.

Chapter 10 of the FAA Airplane Flying Handbook:

https://www.faa.gov/regulations_policies/handbooks_manuals/aviation/airplane_handbook/media/12_afh_ch10.pdf

[14] Airplane Flying Handbook (FAA-H-8083-3C) (full book)

Industry-specific training

Standard operating procedures for an individual industry performance must be developed and are outside the scope of this book. SOPs are an ongoing process, and this is where specific details should develop for the particular needs of a company. Here are some of the industry specifics to consider. Analyze your company and add what specifics you need.

- Agriculture
- Gas and oil inspection
- Mapping
- Real Estate
- SAR
- Thermal inspection
- 3D modeling
- Transmission and Distribution
- Wind turbine inspection
- Additional notes and information:

Additional notes and information:

Safety and Risk Mitigation

Safety and risk mitigation involve being aware of human factors and emergency planning. Human factors and risk management are an intricate part of safety. I recommend having a copy of the Pilot's Handbook of Aeronautical Knowledge and the FAA Risk Management Handbook as a resource. The information provided will elevate your standards to a higher level of aviation.

In the FAA Pilot's Handbook of Aeronautical Knowledge, [15] Chapter 2trave addresses aeronautical decision-making. https://www.faa.gov/regulations_policies/handbooks_manuals/aviation/phak/media/04_phak_ch2.pdf

A Safety Management System (SMS) for commercial operations includes Aeronautical Decision Making (ADM) and Crew Resource Management (CRM).

- ADM = Aeronautical Decision Making. "Decision-making in a unique environment—aviation. It is a systematic approach to the mental process used by pilots to consistently determine the best course of action in response to a given set of circumstances. It is what a pilot intends to do based on the latest information he or she has." (U.S. Department of Transportation. Federal Aviation Administration, Flight Standards Service, 2016, p. 2.1).[16]

- CRM = Crew Resource Management. "Training for flight crews is focused on the effective use of all available resources: human resources, hardware, and information supporting ADM to facilitate crew cooperation and improve decision-making." (U.S. Department of Transportation. Federal Aviation Administration, Flight Standards Service, 2016, p. 2.2).[17]
    - Delegate tasks, address hazards, communication procedures, be open to questions/concerns of all crew members.
    - All crew members should be mindful of physical conditions: Hyperventilation, Dehydration, Heatstroke, Hypothermia, Stress, Fatigue, and Chronic Stress.

The Pilot's Handbook of Aeronautical Knowledge also includes identifying the hazardous attitudes of team members as a form of mitigating risk.

- Five Hazardous Attitudes = **I**mpulsive, **M**achoism, **A**nti-Authority, **I**nvulnerable, **R**esignation."[18] (U.S. Department of Transportation. Federal Aviation Administration, Flight Standards Service, 2016, p. 2.5).
    - I find it helpful to remember "**IM AIR.**"

---

[15] https://www.faa.gov/regulations_policies/handbooks_manuals/aviation/phak/

[16] (U.S. Department of Transportation. Federal Aviation Administration, Flight Standards Service, 2016, p. 2.1).

[17] (U.S. Department of Transportation. Federal Aviation Administration, Flight Standards Service, 2016, p. 2.2).

[18] (U.S. Department of Transportation. Federal Aviation Administration, Flight Standards Service, 2016, p. 2.5).

The Risk Management Handbook[19] provides methods for defining and evaluating risk. Utilize these methods to enhance situational awareness and to detect and avoid the risk. They are designed for traditional aviation, and they are essential for uncrewed aircraft systems as well.

FAA Risk Management Handbook acronyms that are used by the aviation industry include: (U.S. Department of Transportation, Federal Aviation Administration, Flight Standards Service, 2009, pp. 3.2, 5.3, 5.10, 6.8).

- PAVE = **P**ilot, **A**ircraft, en**V**ironment, **E**xternal pressures.[20]
- IMSAFE= **I**llness, **M**edication, **S**tress, **A**lcohol, **F**atigue, **E**motion.[21]
- Decide = **D**etect, **E**stimate, **C**hoose, **I**dentify, **D**o, **E**valuate.[22]
- 3-P = **P**erceive, **P**rocess, **P**erform.[23]
- 5-P = **P**lan, **P**ilot, **P**lane, **P**assengers (**P**ayload), **P**rogram.[24]

Having a plan in place will help reduce the risk and eliminate confusion. Include your physical needs when considering your safety and risk mitigation. Drink plenty of fluids and get plenty of rest before a mission.

Additional notes and information:

____

____

____

____

[19] https://www.faa.gov/regulations_policies/handbooks_manuals/aviation/media/faa-h-8083-2.pdf
[20] (U.S. Department of Transportation, Federal Aviation Administration, Flight Standards Service, 2009, p. 3.2).
[21] (U.S. Department of Transportation, Federal Aviation Administration, Flight Standards Service, 2009, p. 3.2).
[22] (U.S. Department of Transportation, Federal Aviation Administration, Flight Standards Service, 2009, p. 5.3).
[23] (U.S. Department of Transportation, Federal Aviation Administration, Flight Standards Service, 2009, p. 5.10).
[24] (U.S. Department of Transportation, Federal Aviation Administration, Flight Standards Service, 2009, p. 6.8).

Methods for evaluating safety

The "Dirty Dozen"[25] are twelve common errors related to human factors. It is an aviation term for the risks that may be faced during an operation. The concept was originated by an employee with Transport Canada[26] and they are widely used throughout the aviation industry.

Special attention should be given to the aircraft's battery levels or fuel, pilot fatigue, improper hardware, and the condition of the propellers.

The "Dirty Dozen" are used for traditional aviation and are an excellent resource for uncrewed aircraft operations as well.

1. Lack of communication
2. Distraction
3. Lack of resources
4. Stress
5. Complacency
6. Lack of teamwork
7. Pressure
8. Lack of awareness
9. Lack of knowledge
10. Fatigue
11. Lack of assertiveness
12. Norms

Additional notes and information:

[25] https://skybrary.aero/articles/human-factors-dirty-dozen
[26] http://aviationsafetyblog.asms-pro.com/blog/lets-talk-human-factors-origin-of-dirty-dozen

Radio Etiquette

Some work environments will require radio communication between team members. Therefore, it is essential to conduct communication in a professional manner. Practice using radios and communicating prior to a mission.

- Be prepared when ready to speak.
- Start by identifying yourself.
- Identify whom you are addressing.
- Don't talk over others.
- Keep your replies clear, concise, and short.
- Pause before and after.
- Use the phonetic alphabet. I.e., "Foxtrot, Golf."
- Use common radio language. I.e., "Affirmative, Copy, Stand by, Over."
- Set emergency communication protocols.

Additional notes and information:

Traveling with your aircraft

I have been doing quite a bit of national and international travel with my aircraft, and I want to share some tips from my experiences. Every airport and airline will have different regulations. Big help, right? Take time to research the airline(s) you plan to travel with. For smaller aircraft, there are some standard requirements, including:

- Batteries should be removed from the aircraft.
- Batteries are to be placed in your carry-on bag.
- Each battery is placed in a separate battery safe bag.
- An acceptable number of batteries can be carried on based on their Watt-hours (Wh.) up to 160 Wh.

Another tip I have heard in the industry is to discharge your batteries to 50% and cover the terminals with gentle removable tape to reduce the possibility of coming in contact with metal or other objects.

Additional notes and information:

Standard procedures

Pre-Flight Mission Planning

You may have found that flying the aircraft is the "fun" part. However, flying is a small part of the full scope of a mission. Pre-flight planning is often considered the tedious side of a job. This is where your professionalism and concern for safety starts to shine.

Pre-flight and mission planning is mandatory and a crucial step in elevating your standards. Therefore, as a professional, mission planning preparation are expected to be followed, checked, re-checked, and documented. Create several copies of your mission planning for each operation. Copies of documentation and records are kept at the office, electronically, and a paper copy is on hand at the mission site. If you are requested by the FAA, public safety officials, and other government agencies to provide your documentation, you will have a physical copy if your electronics are not available.

Additional notes and information:

Mission planning

Mission planning starts before confirming a job with a client. There may be unknown factors that need to be discussed before taking on the job. Document and record the following information. Each item may be addressed through word documentation, screenshots, images, links, and voice recordings. All items will be saved in the customer/client documentation file. One or more online apps may be used to obtain the information. LAANC, Notice to Air Missions (NOTAM), Temporary Flight Restriction (TFR), and weather are available through multiple resources.

Additional notes and information:

Mission planning preparation list:

- Equipment ready
- Location confirmed
- Pilot/crew/general
- Airports nearby
- Authorizations
- Client needs
- Weather
- Ground station
- Traffic
- TFR
- Risk assesses
- Obstructions
- NOTAM
- Emergency
- Other

Concerns to address

- Permission to fly?
  - Research the area:
    - What is the airspace and nearby airports?
    - Are there NOTAMs and TFRs?
    - Do you need FAA authorization?
    - Do you have the landowners' permission?
    - Are there Federal, State, or local uncrewed aircraft policies in place?
    - Will unlocking aircraft geofence be needed?
- Risk Assessment, Hazards, Privacy & Nuisance
  - What are the hazards?
    - People, animals, schools, trees, powerlines, etc.
  - Are there privacy and nuisance issues that need to be mitigated?
- Weather
  - Current and forecasted weather.
  - Traditional weather patterns during that time?

Mission planning resources

Identity the airspace and Authorizations needed:

- UAS Facility Map. https://faa.maps.arcgis.com/
- LAANC.

  Airspace Link, Aloft, Altitude Angle, Open Sky, Project Wing, Skyward are examples. All resources are available to be used for an airspace authorization. The complete list is available on the FAA website.
- SkyVector. https://skyvector.com/

Airport Information:

- Airspace Link, Aloft, or other LAANC apps.

AirNav. https://www.airnav.com

- SkyVector. https://skyvector.com/
- Nearest Airport. Identify the nearest airport(s) to your area of operation. Use the International Civil Aviation Organization (ICAO) identifier if available.
- Radio frequencies to be monitored: State the frequencies you will monitor for your operation.
- Identify the air traffic patterns of the nearby airports.

Check TFRs and NOTAMs:

- TFRs. https://tfr.faa.gov
- NOTAMs. https://pilotweb.nas.faa/PilotWeb

Map your area: Address hazards and address permission needed.

- Google Maps, Google Earth, or software of your choice.
- Define your parameters.
- Unlock aircraft geofence if needed.

Weather:

- UAV Forecast (available as an app also). https://www.uavforecast.com/
- Aviation Weather app (Apple), or Avia Weather app (Google Play)
- 1800wxbrief Flight Services (Weather & flight planning). https://www.1800wxbrief.com/Website/#!/
- National Oceanic & Atmospheric Administration (NOAA). https://www.noaa.gov/weather

Additional mission planning considerations

Instructions regarding obtaining and the information required for FAA waivers are available online and are outside the scope of this book.[27]

State which of the following FAA waivers you will require (if any):[28] (Federal Aviation Administration (FAA) and Office of the Secretary of Transportation (OST), 2021, pp. title-14/chapter-I/subchapter-F/part-107/subpart-E).

- Flying from a moving vehicle or aircraft, not in a sparsely populated area (107.25).
- Flying at night (107.29).
- Flying beyond the pilot's visual line-of-sight (107.31).
- Flying multiple aircraft with only one pilot (107.35).
- Flying directly over a person or people (107.39).
- Flying near airports / in controlled airspace (107.41).
- Flying above 400 feet (107.51B).

Earliest Date/Time Frame of Operation:

- Specific date(s) and both local/UTC time frames (you must consider waiver processing or the use of LAANC to indicate the earliest likely date of operations if required).

Frequency of Operation:

- Select (approximately) how frequently you plan to operate/how many flights you plan to complete.

Proposed Minimum and Maximum Flight Altitude:

- Using online resources, identify potential hazards for Minimum Obstacle Clearance (MOC).
- Provide the maximum altitude AGL at which the operation will be conducted.

Client-specific:

- Equipment needed.
- Aerial shot list.
- Data collected, deliverables, and analysis.
- Insurance.

[27] https://www.faa.gov/uas/commercial_operators/part_107_waivers/media/waiver_application_instructions.pdf
[28] https://www.ecfr.gov/current/title-14/chapter-I/subchapter-F/part-107/subpart-E

Mission Planning Checklist

Mission Planning Checklist					
Permission	Check/ Fill-in	Documents	Check/ Fill-in	Hazards	Check/ Fill-in
Airspace		Airports		Privacy	
FAA Authorize #		Hospitals		Schools	
Client		Aircraft Registration		Terrain	
NOTAMs / TFR		Aircraft Maint. Log		Trees	
Aircraft Unlock		Pilot Cert		Power Lines	
Weather Cond.		Pilot Log		People	
Local Policies				Animals	
Landowner Permission					

Figure 9 - Mission planning checklist.

Risk Assessment

Evaluate risk and hazards by numbers system. Assume the risk as individual hazards and as a whole. Place the number that is represented in the column.
1= No Factor, 2= Consideration, 3= Use Caution, 4= Potential, 5= High Risk.

Risk Evaluation	No Factor	Considerations	Use Caution	Potential	High Risk	Go/ No Go
Hazard	1	2	3	4	5	Total Value
Aircraft / Airport						0
Weather						0
Terrain						0
Location						0
Trees						0
Schools						0
People						0
Traffic						0
Total Risk Value	0	0	0	0	0	0

Figure 10 - Risk assessment.

Client-specific

Talk with your client and discuss what their vision is for the project or training session. For example, are you working with sensitive data that needs to be handled in a particular manner? Are you filming for a client, and they have an asset they know they want to market? This is also a time for you to share your expertise. If you have a vision of a shot, share it with them.

Occasionally you may be asked to perform something that is not within your skill set or is allowed within the FAA regulations. You are the professional. Don't be afraid to let them know. Provide a solution by coming up with alternative flights that will accomplish what they are looking for.

Have your documentation, flight records, and pilot logs available. You will want to have a copy of your insurance. It is not uncommon for clients to require being added as an additional insured party item on your policy. Secure in writing the required coverages and coverage limits before you set out on your mission.

Know in advance how the data, media footage, is to be delivered. There are various delivery methods, and your client may have specific requirements. Also, remember that not everyone has the same technical skills or computer power to handle information the same way you do. Make a plan for what is needed to accomplish the task.

This is a good opportunity to share a story I have from lessons learned. I drove eight hours to film with a friend on a specific project. When I arrived, I was so overwhelmed with the vast beauty that I couldn't wait to just start flying. Well, I captured amazing footage, but I left the location thinking of all the great shots I didn't get. For example, I always like to get a 3D panorama of an area, and I didn't get one. Had I arrived with my shot list, I would have captured my essential shots and more.

Secure a shot list from your client and develop your preferred shot list for the project. Define the specific flights and camera angles you want to accomplish before arriving and have them written down. This plan may change when you get to a job site but start with a vision.

Additional notes and information:

Client Information Checklist

Client Information Checklist				
Client Specific	Check/ Fill-in	Client Specific	Operation	Check/ Fill-in
Name			Data Collection	
Location			Data Delivered	
Date				
Time				
Mission Focus				
Documents Obtained				
Film Shot List				
Equipment Checklist				

Figure 11 - Client information checklist.

Film Shot List

Film Shot List			
Shot List	Check/Fill-in	Notes	Date
Pan L/R			
Reveal			
Tracking L/R			
POI			
Nadir			
Dolly In/Out			
3D			

Figure 12 - Film shot list.

Pre-Flight Preparation

Pre-flight preparation is imperative to the success of a mission. Arrive prepared for the mission with the proper equipment and personal protection equipment necessary for the completion of the mission. Secure from your client any personal protection equipment requirements. Also, determine if there are any local government or legal requirements.

Personal Protection Equipment

Personal Protection Equipment (PPE) is to assist in preventing accidents and injuries in the workplace environment. Take measures to minimize exposure of personnel to hazardous situations and materials. Personal protection equipment is an essential barrier between workers and the environment. Clothing and equipment that are worn to prevent or minimize hazards to personnel are critical. PPE will maintain safety and quality in your workplace. PPE includes, but is not limited to, a hardhat, gloves, steel-toed shoes, safety glasses, a high visibility jacket or vest, a hearing protection device, sunscreen, and insect repellent.

Each crew member is responsible for their PPE unless it is communicated that PPE will be provided. Utilize the checklist and add additional items that are required per each mission.

Additional notes and information:

____

____

____

____

PPE Checklist

PPE Checklist			
PPE	Check/Fill-in	Addition PPE	Check/Fill-in
Water / Food		Boots	
Medications		Bug spray	
High Vis Vest		Face Mask	
Hard Hat		Sunscreen	
Long pants/gators		Eye Protection	
Gloves			
Emergency kit			

Figure 13 - PPE checklist.

Equipment

Ensuring you have all your equipment prepared and your vehicle is properly loaded is an essential part of elevating your standards. There is nothing worse than arriving on a job site and finding that you have left something out. Your entire job can be on hold if you have forgotten something as small as an SD card or a tablet cable, and this is a reflection of your professionalism. Developing a checklist for your equipment will help to ensure you have all the necessary items when you arrive on-site.

Inventory checklists for equipment checkout are completed before each mission. Each item is inspected for its integrity, and the software is up to date. Batteries, remote controls, and accessories are to be fully charged. Enter the name of the crew member responsible for checking out items.

Equipment Checklist

Equipment Checklist		Checked Out By:		Date:	
Equipment	Check/ Fill-in	Equipment	Check/ Fill-in	Equipment	Check/ Fill-in
Aircraft		Cones		Safety Gear (PPE)	
Batteries (qty)		Launch/Land Pad		Shovel	
SD Cards (qty)		Table		Emergency Kit	
Controller (s)		Chair			
Laptop		Shade			
Internet		Auto Fueled			
Tablet		Auto Loaded			
External Monitor					
Portable charge					
Documents					

Figure 14 - Equipment checklist.

On-Site Procedures

Have your checklists ready and easily available when you arrive. Put them on a clipboard, and you may want to have a clip to hold the bottom down to keep them from flying up in the wind. Develop good habits, don't become complacent. When flying a large mission with multiple flights, I recommend exchanging SD cards between flights. Should there be an unforeseen incident, you will not have lost your prior flight data.

Known policy procedures for on-site procedures maximize safety, eliminate crew member confusion, and will elevate efficiency. Maintaining the highest level of safety is accomplished by following set policies and procedures in your ground and flight operations. Communication and utilizing checklists elevate efficiency.

- Safety Brief:
    - Reconfirm eliminating risks, avoid hazards, and MOC.
    - Confirm checklists.
        - Mission, crew member responsibilities, and data capture and delivery.
    - Identify battery life distance aircraft has potential to fly.
- Setup launch/land command station and the aircraft.
- Unlock the aircraft geofence if required.
- Call ATC if required.
- Execute mission.
- Verify collected data before leaving the site.

Additional notes and information:

On-site operations checklist

Name: ______________________________ **Date/Time:** __________________________________

Safety brief tailboard

Before the flight, confirm checklists, mission, responsibilities, and data capture with the crew members.

Verbally communicate and confirm with the VO and Crew their objectives, roles, and responsibilities. Discuss the hazards and emergency procedures before arming the aircraft and setting up a command station.

- ☐ Confirm risk and hazard assessment.
- ☐ Confirm flight procedure, mission objectives, minimum obstacle clearance (MOC), checklists.
- ☐ Confirm pre-flight planning data, ATC, hospital, etc.
- ☐ Confirm the weather minimums.
- ☐ Confirm ADM and emergency procedures.
- ☐ Confirm battery life flight distance.
- ☐ Confirm flight area safe to set up.

Set up launch site flight line, command station, aircraft

Flight line setup and command station.

- ☐ Launch/land pad placed "North" and approx. 15' or more away from the pilot and crew.
- ☐ Mark area with cones to allow safe launch/land for the appropriate size of the aircraft. Approx. 10'x10'.
- ☐ Confirm radio with ATC frequency (power on) if in use.

Aircraft.

- ☐ Confirm the structure is fit for operation.
- ☐ FAA number labeled.
- ☐ Propellers are secure.
- ☐ Battery(s) are secure.
- ☐ Camera(s) are secure.
- ☐ SD card in place.
- ☐ Verify with Visual Observer the aircraft is ready for operation by reconfirming steps.

Operations.

- ☐ Power on the remote control.
- ☐ Power on the tablet, start the app. being used to control the aircraft.
- ☐ Power on the aircraft.
- ☐ Verify the telemetry (clockwise procedure) to confirm the data.
- ☐ Verify battery levels.
- ☐ Verify Remote ID is functioning.
- ☐ Set and confirm the Return To Home (RTH) elevation, the RTH position, and maximum flight altitude.
- ☐ Unlock aircraft geofencing if needed. To unlock DJI: http://Dji.flysafe.com.
- ☐ Document the aircraft and battery information.

Launch aircraft.

- ☐ Sterile cockpit.
- ☐ Announce "clear on the flight deck."
- ☐ Crew replies "clear" to confirm all crew members are aware the aircraft is active.
- ☐ Elevate the aircraft approximately 10 - 12'.
- ☐ Visually inspect the disposition of aircraft (responsiveness to weather, wind tolerance, etc.).
- ☐ Check aircraft telemetry to confirm the analysis.
- ☐ Maneuver approximately 15' away after takeoff test stick controls. Test each position pitch, yaw (rotate), roll, climb, and descend.
- ☐ Test the camera movement.
- ☐ Rotate and capture an image of the flight line and surrounding area.

Flight.

- ☐ Reconfirm MOC if and adjust if needed.
- ☐ Perform mission.
- ☐ Confirm mission complete.

Landing aircraft.

- ☐ If needed, capture an image to distinguish the end of the mission: blue sky or landing pad.
- ☐ Clear flight deck - Confirm clear to land.
- ☐ Confirm camera up position before landing.
- ☐ Reconfirm "clear to land."
- ☐ Disarm aircraft motors.
- ☐ Verify the data collected.
- ☐ Shut down the aircraft.
- ☐ Shut down the remote control.

Post-flight and deliverables.

- ☐ Check data before leaving the site!
- ☐ If the client requests, pass data directly to the client.
- ☐ If processing data, verify through the software of choice to confirm the integrity of data collected.
- ☐ Before leaving the site, double-check the area to remove trash or items forgotten.

Figure 15 - On-site operations checklist.

Flight information checklist

Flight information: ____________________ **Client Name:** ____________________

Date: ____________________ Flight Name: ____________________

Long/ Lat: ____________________ Client Name: ____________________

Address: ____________________ Description: ____________________

NEARBY AIRPORTS:

Name & ID: ____________________ ATC Phone #: ____________________

Address: ____________________ Contact Person: ____________________

NEARBY HOSPITAL:

Name: ____________________ Phone #: ____________________

Address: ____________________ Notes: ____________________

CREW:

Pilot / VO: ____________________ Pilot / VO: ____________________

Name: ____________________ Name: ____________________

107 Cert: ____________________ Cert/Training: ____________________

AIRCRAFT

Aircraft 1: ____________________ Aircraft 2: ____________________

Make/model: ____________________ Make/model: ____________________

Reg #: ____________________ Reg #: ____________________

WEATHER:

Wind Direction/Speed: ____________________ KP: ____________________

Temperature: ____________________ Time Taken: ____________________

TFR / NOTAMs:

TFRs: Yes_____ NOT in Vicinity _____ NOTAMs: Yes _____ No ______

SAFETY EVALUATION: 1 - 5 (1= low, 5=severe)

Weather_____ Utility lines_____ Environment_____ People_____ Other_____ Total_____

Pilot flight tracking						
Pilot name	Battery #	Volt & % Start	Volt & % End	Pilot Start Time	Pilot End Time	Total Flight time

Figure 16 - Flight information.

In-Flight operations

During planned flight missions, the pilot must comply with the FAA regulations unless otherwise authorized. The pilot must be aware of hazards, obstacles, and other aircraft. A visual observer must be utilized when required. Uncrewed aircraft operations are not to be conducted in a reckless manner and must yield the right-of-way to other aircraft. It is the responsibility of the Remote Pilot in Command (RPIC) to ensure the aircraft and crew are compliant.

- Not to exceed 400' AGL, unless flown within a radius of 400' from a structure.
- Operate within visual line-of-sight.
- Require minimum visibility of three (3) statute miles.
- Provide Anti-collision lights 30 minutes before and after sunset or sunrise (civil twilight).
- 500' below a cloud, 2000' Horizontal from clouds.
- May not consume alcohol in the preceding eight (8) hours or have a blood alcohol concentration of .04 Blood Alcohol Concentration (BAC).
- No flights are faster than a ground speed of 87 knots (100 mph).
- Operations from a moving land or water vehicle are in a sparsely populated area.
- May operate only one aircraft at a time.
- At this time, the remote pilot may not operate an uncrewed aircraft over another person unless compliant with new rules of Remote ID and sUAS categories. Otherwise, person/people are under a safe cover, a stationary vehicle, or a protective structure.

Post-Flight and Deliverables

Verify your data before you leave the site! First, make sure it is clear, and the exposure is what you need. Then, double-check to verify you have all the required footage.

Pack it in, pack it out. Be respectful of your surroundings. It is not enough to just pack up your belongings. Remove any trace that you were there. For example, if you have left huge tire imprints, smooth them out. Pick up your trash and close the gates. Go over the equipment list as a reminder of what you brought with you to the site.

Have a documented system that shows the deliverables have been provided to the client. Make a note of the date the deliverables were passed off.

- Check data before leaving the site!
- If the client requests, pass it directly to the client.
- If processing data, verify through the software of choice to confirm the integrity of data collected.
- Before leaving the site, double-check the area to remove trash or items forgotten.

Additional notes and information:

Debrief

Debriefing is often a forgotten step. Yet, this is a great way to improve your flights and verify the success of the mission. A debrief will elevate standards to improve future missions and company protocols. Whether the mission was solo, with a VO, or with a large team, have a record of completion for your files and build standards upon them.

The pilot and crew will conduct a debriefing upon completion of the mission. Summarize the success of the mission. Include pertinent information regarding the crew, equipment, and location.

Additional notes and information:

Mission Debrief

The pilot and crew will conduct a debriefing upon completion of the mission. Summarize the success of the mission. Include pertinent information regarding the crew, equipment, and location.

Pilot Name: ______________________________ Date/Time: ___________________________

Was the mission successful?
- ☐ Yes
- ☐ No
- ☐ Explain: ____

Was the necessary data captured?
- ☐ Yes
- ☐ No
- ☐ Explain: ____

Did the equipment function properly?
- ☐ Yes
- ☐ No
- ☐ Explain: ____

Were there issues with the mission:
- ☐ Yes
- ☐ No
- ☐ Explain: ____

How can we mitigate issues?
- ☐ Explain: ____

Was the crew focused and alert?
- ☐ Yes
- ☐ No
- ☐ Explain: ____

Can this mission improve?
- ☐ Yes
- ☐ No
- ☐ Explain: ____

Figure 17 - Mission debrief

Data Folder Setup

Have you ever downloaded footage and wondered where it went? Did it get lost in the computer abyss? Developing a systematic way of importing and storing your data will help alleviate this problem. I have made a template blank file folder on my computer with the pertinent directories I like to use. Then I can simply copy/paste and rename the folder for the project I am working on.

Systematically storing data will save time, money, and frustration. Most software will follow a specific path to the information needed. Such as images. Once they are moved, the software will need to be directed to the new location or will not be able to process the data. Do not use a long directory path, not more than thirteen (13) characters, no special characters. For example, C://desi/103121…

Additional notes and information:

Example of an electronic file folder:

- Project name and date
- Audio
- Documents
- Exported
- Graphics
- Media Footage
  - Still photos
  - Video
- Data
  - Boundary KML
  - Control GCP
  - Images
  - Grid
  - Orbit
  - Project software specific
- Video project

Care and Maintenance

Why wouldn't you take care of and maintain your equipment? It's your money maker. When you arrive on-site, you want to know everything is going to operate correctly and safely for you. Elevate your standards by maintaining your assets. Your professionalism will shine when you arrive prepared and ready for the operation.

This goes beyond maintaining your aircraft. Keep all your equipment updated and in good working condition. For example, replace frayed wires or tablet cables that show visible damage.

Keep your equipment in a dust-free environment as much as possible. This is a tough one for me since I spend a good part of my time filming off-road in dusty desert environments. I bring a large landing pad which is very helpful with the amount of dust the Inspire 2 creates.

Develop a documentation system for the updates and repairs you conduct on your aircraft. Keep these records available for the FAA also.

Aircraft Maintenance
Visual inspection is performed before and after each mission. Report and document any necessary information. Include firmware updates as well as aircraft hull maintenance that has been completed in your documentation.

Inspection of aircraft will include but is not limited to:

- All sides of the aircraft.
- Landing gear.
- Screws and attachments.
- Propellers.
- Motors.
- Gimbals.
- Wiring.

Additional notes and information:

Aircraft Maintenance Log

Aircraft Maint. Log							
Date	Aircraft ID	Aircraft Number	Personnel	Maintenance Performed	Firmware #	Notes	Result

Figure 18 - Aircraft maintenance log.

Batteries

Lithium polymer (LiPo) batteries can be dangerous if mishandled during charging, use, storage, and transport. Proper care of aircraft batteries will maximize your flight time. Care and maintenance of aircraft batteries are a critical part of pre-flight, post-flight, and maintenance procedures.

DO:

- Refer to the manufacturer's guidelines for the safe handling of batteries.
- Only use the charger designed for the battery. Use of an incorrect charger can cause damage or injury.
- Use a LiPo safe bag to charge the battery safely.
- Cycle the use of batteries in a rotation to allow one battery not to be used more than the others.
- Check the voltage before and after you fly.
- Document flight times with batteries.
- Have a bag of sand on hand for battery emergencies.
- Take them to a battery recycling location to dispose of them.

DON'T:

- Do not use if any type of damage, leak, or "puffy" swelling is found on the battery.
- Do not bend, drop, or hit the battery.
- Do not store or transport batteries with metal objects.
- Do not charge a battery inside of the aircraft.
- Do not discard the battery in the trash or fire.
- Do not leave a charging battery unattended.
- Do not charge a battery immediately after flying. Wait and allow time for the battery to cool down.

Storing Batteries:

- Short term – discharge to 60 and 80%, if no flights are planned for one (1) to ten (10) days.
- Long-term – discharge to 40 and 60% if not flights are planned for more than ten (10) days. Batteries should not be left for over three (3) months without being charged.
- Do not store batteries in a vehicle other than for transportation. Keep batteries in the shade or a cool spot while on location.

Additional notes and information:

Care and Maintenance of Aircraft Propellers (Props)

Aircraft propellers are to be inspected regularly and before each flight. Inspect each propeller for nicks or cracks. Visually look for signs of weakness in the integrity of the propeller. They should be stored in their case or in a manner not be bent or damaged during transportation.

Propellers are essential to the flight of an aircraft. Do not use propellers that appear to be damaged, broken, or warped.

Additional notes and information:

Emergency procedures

Emergencies

Elevate your standards by planning and training for proper emergency procedures. Nobody wants an emergency situation, but having a plan will help to keep a crisis from rapidly escalating. Identifying various emergencies ahead of time will help put a plan into action. It is also good to walk or talk through an emergency situation and go through the motions of action.

This is a small list of the possible emergency situations you may have during your uncrewed aircraft operations. Continue to build out scenario-based emergencies and responses with your team. For example, practicing flights in "Atti-mode" will help you to confidently fly the aircraft if there is a loss of GPS.

Company standards are in place to identify crew member roles during an emergency. Each situation will vary, but planning will eliminate confusion. An emergency can range in severity and may need to be reported to a higher authority.

It is mandatory that a serious uncrewed aircraft incident is reported to the FAA and the National Transportation Safety Board (NTSB) within ten (10) days of the occurrence.[29] (Federal Aviation Administration (FAA) and Office of the Secretary of Transportation (OST), 2021, pp. title-14/chapter-I/subchapter-F/part-107/Subpart A - General).

Utilize the FAA DroneZone portal for reporting accidents. Also, the National Aeronautics and Space Administration (NASA) has a website UAS Aviation Safety Reporting System (ASRS). This site is used for close calls, hazards, violations, and safety-related incidents.

FAA accident reporting information:
https://www.faa.gov/documentlibrary/media/order/faa_order_8020.11d.pdf

NASA UAS reporting link:
https://asrs.arc.nasa.gov/uassafety.html

Follow NTSB guidelines:
https://www.ntsb.gov/Pages/Report.aspx
To file a report with the NTSB, contact the NTSB 24-hour Response Operations Center (ROC) at 844-373-9922. A written follow-up may be required.

[29] Federal Aviation Administration (FAA) and Office of the Secretary of Transportation (OST), 2021, pp. title-14/chapter-I/subchapter-F/part-107/Subpart A - General/107.9 Accident reporting.

Emergency responses

To properly respond to an emergency, you must be prepared ahead of time. This requires training and directives prior to an incident.

1 – Training: Be prepared as much as possible before an incident.
2 – Outlined directives: For pilot, crew, and post responses.
3 – Documentation: Being cautious of sensitive information, document site with photographs and written descriptions statements.
4 - Emergency information: Numbers, names, and links to handle the emergency.

Emergency situations and severity will vary and will not require the same responses. A temporary loss of GPS signal will not require the same response as a battery fire.

In the case of loss of GPS, RPIC must not over-rely on the aircraft's automation and be prepared to operate aircraft manually. A loss of GPS may happen during a flight or be due to a GPS anomaly. By checking NOTAMs prior to a mission, you may become aware of GPS anomalies. Advanced training with flights in "Atti" mode and the use of flight termination procedures are imperative.

Aborting or terminating a flight will intentionally abort the mission process by the RPIC. During your pre-flight planning, identify an emergency landing area and abort procedures. When descending an uncrewed aircraft at a rapid rate, avoid prop-wash (vortex ring state) by descending in a spiral flight pattern.

Battery fires or a thermal runaway are hazards to be concerned with regarding uncrewed aircraft. Batteries can be very volatile. Check batteries before and after your flights. Do NOT use any damaged or puffy batteries. Care and maintenance of your batteries are crucial. Do not use water to extinguish a lithium battery fire. Cover the battery with sand if a Class D extinguisher is not available.

Incident scenario: Lost link

Loss of transmission communication to/from aircraft.
A lost link is an interruption or loss of the upward or downward control link between the control station and the aircraft. This is a common occurrence with uncrewed aircraft and needs to be addressed. Loss of transmission communication will prevent control of the aircraft if the connection is not reestablished. The lost link interruption is usually brief. The remote control should be facing the aircraft and confirm there is no structure or tree blocking the signal from the remote control to the aircraft.

Pilot Response:

- Verbalize with the crew immediately that the link has been lost.
- Remain airborne and attempt to reestablish the link by waiting in a hold position for approximately 5 – 10 seconds.
- If not responding, toggle the control switch from "Atti" or "Smart" mode and back to "Position" to reset the connection.
- If the lost link is between the aircraft and the controller is not reestablished, utilize Return to Home (RTH) and/or Autoland feature.
- Commence and verbalize with the crew planned procedures that have been previously discussed during mission planning for the emergency landing area.

Crew Response:

- Everyone is on alert.
- Clear the landing area.
- Identify aircraft location.
- Verbalize between crew members action taken.

Post Incident Action:

- Discuss possible causes for the link disruption such as equipment failure, pilot error, magnetic or WIFI interference.
- Take action to avoid future incidents.
- Write a log report of the incident with specific information regarding the pilot information, and the aircraft make/model, the aircraft FAA number, and the incident. (Provided document).
- Write a company report after an incident.
- Write a report for the FAA and NTSB if required.
- Report the incident to local law enforcement if required.

Incident scenario: Wildlife/bird strike or animal interference

Be very aware of nearby animals and wildlife. Animals will feel threatened by people and by aircraft in their space. This may cause the animal or wildlife to intentionally attack people or the aircraft. Animals and wildlife may also be disturbed by the noise of the aircraft. Stay alert to your surroundings.

Pilot Response:

- Verbalize to the crew all wildlife/bird or animal activity in the area.
- If the aircraft is under attack from other animals or birds:
  - Wildlife/Birds:
    - Depending on the type of bird, hold a position to wait for birds to pass by and allow them time to settle or avoid.
    - Or climb in elevation to avoid an attack from above.
  - Ground Animals:
    - Elevate to avoid animals on the ground and to remove the disturbance.
- Remove the aircraft from the flight zone of the birds and animals.
- If a strike occurred, land immediately.
- Look for the aircraft, any damage caused, or injured bird or animal after the incident.
- Assess the bird or animal, stay out of harm's way. Contact Wildlife Fish and Game, a veterinarian, or the park ranger.

Crew Responses:

- Assist the pilot with visuals of birds, animals, and the aircraft.
- Keep ground animals away from the pilot (e.g., dogs).
- Verbalize when Wildlife/Birds are present.
- Once the aircraft is safely landed, assess the bird or animal and get medical attention necessary.
- Being cautious of sensitive information, document site with photographs and written descriptions statements.

Post-incident action:

- Document incident information using incident report form.
- Document wildlife, pilot and number, aircraft make/model, aircraft FAA number, date, time, weather, and location of the incident.
- Report to authorities and owner.
- FAA Document AC 150/5200-32B regarding wildlife strikes: https://www.faa.gov/documentlibrary/media/advisory_circular/150_5200_32b.pdf

Incident scenario: Crew member emergency

An emergency may occur at any time during the mission. These could be from one or more scenarios ranging from a pilot not being prepared for work duties to an impromptu situation during the mission. A crew member emergency could be a result of a physical or physiological condition preexisting or brought on by a current situation.

1. Crew member not arriving ready to perform: IMSAFE.
2. A physical condition that sets in during a mission: Health and safety due to harsh environmental issues.
3. An accident that has occurred during a mission: Environment and equipment hazard.
4. Pilot or crew: Not appropriately trained for the mission.

In a situation of a dire emergency call, 911 immediately. (As part of pre-flight planning, local hospitals and numbers are readily available).

Pilot response:

1. Crew member has not arrived ready to perform required tasks.
    - Notify the lead team pilot or the crew member they are not ready to fulfill their duties.
    - Utilize the FAA "IMSAFE" and assess if the pilot is fit for flight.
    - Do not fly if conditions exist.
    - The pilot may be assigned to other non-taxing duties if deemed fit.

2. Physical or physiological incident condition occurs during a mission.
    - STOP flying the aircraft immediately if there are any signs of hyperventilation, dehydration, heatstroke, hypothermia, stress, and fatigue.
    - Seek medical attention and treatment.
    - Notify team lead of occurrence.

Crew Response:

- Clear the area for treatment.
- Assists the team with seeking or providing medical treatment.
- Post-incident action:
- Document incident information: pilot and number, aircraft and number, date, time, and location of the incident.
- Write a report of the incident and the severity. Being cautious of sensitive information, document site with photographs and written descriptions statements.

Fly away / aircraft collision

EXTREMELY DANGEROUS. Do not take this lightly. A flyaway can occur when the aircraft and the controller are not linked and communicating. In this condition, the user cannot control the aircraft.

Pilot response:

- Immediately notify crew members.
- Attempt to regain connection with the aircraft. This can be done through
- "Atti" mode or Smart mode or Return to Home function.
- If the aircraft flyaway continues, note aircraft battery life, height, speed, and heading.
- Inform the police via 911 and ATC at the number that was documented on the pre-flight checklist.
- Maintain Visual Line of Sight (VLOS) with the aircraft for as long as possible.
- Notify team lead.
- Look for the aircraft, property damage, or injured individuals.

Crew Response:

- All hands-on deck.
- Notify verbally or radio communication of aircraft heading.
- Keep visual on aircraft as long as possible.
- Look for the aircraft, property damage, or injured individuals.
- Assists the team with record-taking and reporting information. Being cautious of sensitive information, document site with photographs and written descriptions statements.

Post-incident actions:

- Prepare information for reporting. Follow-up with the proper authorities, insurance carrier, and injured individuals as needed.
- Document the condition of the aircraft with photographs, written descriptions, and flight logs.
    - Remove or repair the aircraft.
- Write a report for the company records, for the FAA, and the NTSB.
- Submit your report through the FAA DroneZone portal. https://faadronezone.faa.gov/#/
- Submit your report to the NTSB. https://www.ntsb.gov/Pages/Report.aspx
- Report the incident to local law enforcement if required.

Incident Report

In the case of an incident, a company incident report should be written within 24 hours of the incident. This will keep the information fresh, and current should something arise at a later time. Notification to the team lead is reported immediately.

The FAA requires a serious incident to be reported within ten (10) days of the incident. Proper documentation is reported to the FAA if injury or damage is more than $500 fair market value or serious injury. The National Transportation Safety Board (NTSB) must also be notified of serious injuries.

Submit your report to the FAA through the DroneZone portal:
https://faadronezone.faa.gov/#/

FAA accident reporting information
https://www.faa.gov/documentlibrary/media/order/faa_order_8020.11d.pdf

Follow NTSB guidelines
https://www.ntsb.gov/Pages/Report.aspx
To file a report with the NTSB, contact the NTSB 24-hour Response Operations Center (ROC) at 844-373-9922. A written follow-up may be required.

Additional notes and information:

Incident Report Form

Date / Time of Incident: ________________ Authority Contacted: __________________
Location of
Incident: ____

Pilot Information:
Full Name PIC _________________________ Pilot Certificate # ____________________
Pilot Phone # __________________________ Pilot Email _________________________

Visual Observer Information:
Full Name VO_________________________
VO Phone #___________________________VO Email___________________________

Aircraft Information:
Aircraft Make & Model__________________ Aircraft FAA #______________________
Identifying Characteristics____

Aircraft Owner Name
Owner Phone #_________________________Owner Email________________________

Weather Conditions_____________________ Other Conditions ___________________

Description of incident: *Include as much description and as many details as possible. Use additional paper if needed.*

____
____
____
____
____
____
____
____
____
____
____

PIC Signature: ___________________________ Date/Time: _______________________

Figure 19 - Incident report form.

This page intentionally left blank.

Conclusion

By implementing the safety protocols, checklists, and resources provided in this book, you are on your way to elevating your uncrewed aircraft operations standards. Safety and documentation are mandatory and cannot be taken lightly. The information within this book is to be used as guidance for elevating your standards. It is your responsibility to utilize the suggested material, do further research, and continually add to your standard operating procedures. Starting your operations manual can be an overwhelming task. With the help of this book and the materials provided, you are developing safety checklists and company standards that will elevate you and your company's professionalism.

Please reach out and contact me if you are creating your SOPs and have any questions. I am on most platforms of social media. Look for Drone Diva® Desi.

LinkedIn: www.linkedin.com/in/desiree-ekstein

Abbreviations

AC = Advisory Circular
ADM = Aeronautical Decision Making
AGL = Above Ground Level
AMA = Academy of Model Aeronautics
ANSI = American National Standards Institute
ASRS = Aviation Safety Reporting System
ASTM = American Society for Testing and Materials
ATC = Air Traffic Control
ATTI= Attitude
AUVSI = Association for Unmanned Vehicle Systems International
AUVSI TOP = AUVSI Trusted Operator Program
BVLOS = Beyond Visual Line of Site
CFR = Code of Federal Regulations (14 CFR Part 107)
COA = Certificate of Authorization
COW = Certificate of Waiver
CRM = Crew Resource Management
CTAF = Common Traffic Advisory Frequency
DOC = Declaration of Compliance
FAA = Federal Aviation Administration
FCC = Federal Communication Commission
FPV = First Person View
FSDO = The Flight Standards District Office
FSS = Flight Service Station
GPS = Global Positioning System
JARUS = Joint Authorities for the Rulemaking of Unmanned Systems
LAANC = Low Altitude Authorization Notification Capability
MOC = Minimum Obstacle Clearance
NAS = National Airspace System
NIST = National Institute of Standards and Technology
NOTAM = Notice to Air Missions
NTSB = National Transportation Safety Board
OMV = Over Moving Vehicle
OOP = Operations Over People
PIC = Pilot In Command
RID = Remote Identification, or Remote ID
RPIC = Remote Pilot in Command
RTH = Return to Home
SFC = Surface
SMS = Safety Management System
SOPs= Standard Operating Procedures
SORA = Specific Operations Risk Assessment
TFR = Temporary Flight Restriction
UAS = Unmanned Aircraft System
sUAS = Small Unmanned Aircraft System
UAV = Unmanned Aerial Vehicle
VLOS = Visual line of sight
VO = Visual Observer

Helpful Links

These are the links at the time of this writing. Some of the links may change with updates and regulation changes.

Academy of Model Aeronautics: http://www.modelaircraft.org/

AirNav: https://www.airnav.com/

Airspace Information:
https://www.faa.gov/regulations_policies/handbooks_manuals/aviation/phak/media/17_phak_ch15.pdf

eCFR: https://www.ecfr.gov/current/title-14/chapter-I/subchapter-F/part-107#part-107

FAA 107 Summary: https://www.faa.gov/uas/media/Part_107_Summary.pdf

FAA Aircraft Flying Handbook: Airplane Flying Handbook (FAA-H-8083-3C) (full book)

FAA DroneZone: https://faadronezone.faa.gov/#/

FAA Risk Management Handbook:
https://www.faa.gov/regulations_policies/handbooks_manuals/aviation/media/faa-h-8083-2.pdf

FAA Pilot's Handbook of Aeronautical Knowledge:
https://www.faa.gov/regulations_policies/handbooks_manuals/aviation/phak/media/pilot_handbook.pdf

FAA Safety Management System: https://www.faa.gov/about/initiatives/sms/

FAA UAS: https://www.faa.gov/uas/

Flying under rules of model aircraft: https://www.faa.gov/uas/faqs/#ffr

METAR and TAF decoding: reference:
https://business.desu.edu/sites/business/files/document/16/metar_and_taf_codes.pdf

NOAA National Weather Service: https://aviationweather.gov/static/help/taf-decode.php#ICAO

TFR: https://tfr.faa.gov/tfr_map_ims/html/

UAS Data Exchange (LAANC):
https://www.faa.gov/uas/programs_partnerships/data_exchange/#approved

UAS Facility Map – The Grid:
https://www.faa.gov/uas/commercial_operators/uas_facility_maps/

Table of Figures

References

Federal Aviation Administration (FAA) and Office of the Secretary of Transportation (OST), D. o. (2021). eCFR.gov. Federal Aviation Administration (FAA) and Office of the Secretary of Transportation (OST), Department of Transportation (DOT).

U.S. Department of Transportation, Federal Aviation Administration, Flight Standards Service. (2009). Risk Management Handbook. In *Risk Management Handbook https://www.faa.gov/regulations_policies/handbooks_manuals/aviation/media/faa-h-8083-2.pdf* (pp. 2.5, 3.2. 5.3, 5.10, 6.8). Produced by the Federal Aviation Administration (FAA) with the assistance of Safety.

U.S. Department of Transportation, Federal Aviation Administration, Flight Standards Service. (2021). Airplane Flying Handbook. In *Airplane Flying Handbook https://www.faa.gov/regulations_policies/handbooks_manuals/aviation/airplane_handbook/media/00_afh_full.pdf* (p. Ch. 10). Published by the United States Department of Transportation, Federal Aviation Administration, Airman.

U.S. Department of Transportation. Federal Aviation Administration, Flight Standards Service. (2016). Pilot's Handbook of Aeronautical Knowledge. In *Pilot's Handbook of Aeronautical Knowledge https://www.faa.gov/regulations_policies/handbooks_manuals/aviation/phak/media/pilot_handbook.pdf* (p. Chapter 2). Published by the United States Department of Transportation, Federal Aviation Administration, Airman.

This page intentionally left blank.

About the author

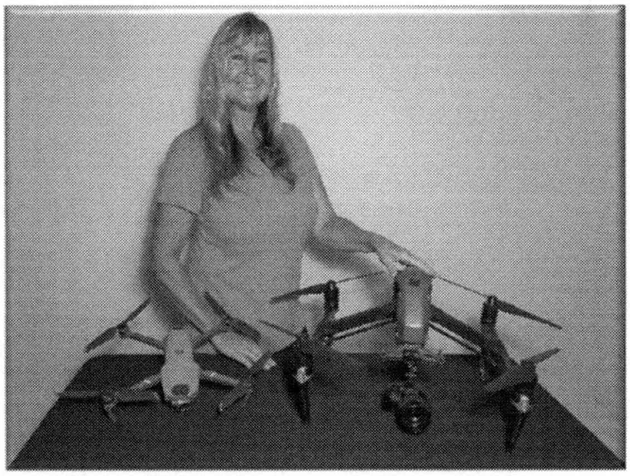

Desiree Ekstein, widely recognized as "Drone Diva Desi," is an accomplished figure in the field of Unmanned Aerial Systems. Her diverse achievements and contributions to the industry have solidified her reputation as a trailblazer and expert in the field.

Desiree is the founder of On The Go Video.biz, an enterprise dedicated to providing training and services related to drone technology, video production, mapping, and 3D models. Notably, she was among the pioneering women to receive the FAA Part 107 Remote Pilot Certificate in 2016, showcasing her commitment to advancing the emerging drone industry.

Desi's expertise extends to various facets of drone operation and instruction, including aerial inspections in infrastructure, agriculture, and construction. She is Pix4D and DroneDeploy certified, with significant experience in using these platforms for data collection and documentation.

As an instructor of UAS, she also contributes to the "Girls Take Flight" STEM program, a young female practicum, providing drone training to future generations. Her qualifications are further exemplified by her AUVSI TOP Level 3 pilot and instructor status. She serves as the UAS Lead FAASTeam Representative and DronePro in San Diego, California, advocating for drone safety.

Desi's influence extends internationally as a public speaker. She has shared her expertise at Africa: Drones and Unmanned Aviation Conference, Drone Summit Ireland, Commercial UAV Expo, and Women in Technology International. As a published author, she continues to educate others about the evolving drone industry, regulations, and its immense potential. Her dedication to safety and training in the industry has earned her induction into the Women in Emerging Aviation Technology Hall of Fame and the FAA Part 107 Trailblazer award.

Desiree Ekstein's remarkable achievements and contributions to the drone industry position her as a respected figure and advocate for both safety and innovation in the field.

This page intentionally left blank.

Made in United States
Troutdale, OR
11/03/2026

56844390R00046